W0007278

Andrew Eborn's
Book of Failure®
From The Octopus TV
Failure® Awards

"If necessity is the mother of invention,
Failure® is the father of success."

By Andrew Eborn

Andrew Eborn's Book of Failure®
By Andrew Eborn

A catalogue record for this book is available from The
British Library
Published by Hope & Plum Publishing
www.hopeandplum.com

ISBN 978-1-8380302-1-6

Text © Andrew Eborn 2019
Cover design © Andrew Eborn 2020
Book under exclusive license to Hope & Plum

Cover design and Photograph of Andrew Eborn
courtesy of Corrado Amitrano

The moral rights of the author have been asserted.

All rights reserved. No part of this publication may be reproduced, stored in a retrieval system or transmitted in any form or by any means, electronic, mechanical, photocopying, recording or otherwise without the prior written consent of the copyright owner and subject to a fully executed license.

DEDICATION

To my fantastic family – thank you for never failing to keep me grounded in this mad, mad world and enabling and encouraging me to fly and explore areas so often taken for granted or ignored.

#QUESTION EVERYTHING

My children are my best productions – EVER!

QUOTES

"The only real mistake is the one from which we learn nothing." Henry Ford

"Failures are finger posts on the road to achievement." C.S. Lewis

"Success is stumbling from failure to failure with no loss of enthusiasm." Winston Churchill

"If necessity is the mother of invention, Failure® is the father of success." Andrew Eborn

ANDREW EBORN PREDICTS ....

For many years Andrew Eborn has empowered companies to face the challenges of changing markets, maximise the return on their rights and assist with the strategic development of their businesses.

As a Member of The Inner Magic Circle with Gold Star as well as a lawyer and strategist, predictions and a deep understanding of human behaviour come with Andrew Eborn's territory – making the impossible possible.

He has successfully predicted major events, developments in technology and entertainment as well as election results with alarming accuracy.

Andrew has always been at the forefront of embracing technology and pioneering developments in entertainment including:

- helping Pioneer to bring Karaoke to the UK
- enabling - through Octopus TV - some of the first live streaming of sporting and other events
- providing the technology backbone for video on demand for newspaper groups and brands enabling them to be their own broadcasters
- revolutionising the distribution of content by facilitating the global digital delivery of files

i

- pioneering immersive entertainment, augmented reality and holograms

The advances in technology and the increase in bandwidth continue to open more and more opportunities.

In Cannes in 2019, in addition to discussing Octopus TV's and Knot the Truth's rich slate of diverse programmes across each genre, Andrew Eborn was engaged by NHK to help present their 8K content to the world.

For the first time in Cannes history, 8K content was shown on a 248 inch screen with 22.2 channel surround sound. Simply stunning!

Andrew also introduced the international premiere of the dramatisation of Nobel Laureate Kazuo Ishiguro's novel *An Artist of the Floating World* in 8K.

Andrew Eborn interviewed several A-list celebrities including Oscar nominee and multi award winning actor Ken Watanabe (star of *The Last Samurai* and a number of other Hollywood films such as *Batman Begins*, the *Godzilla* reboot, *Inception* and the *Transformers* series.) The interview was filmed in 8K for NHK BS8K channel and NHK News. It will be broadcast in December on the 1st anniversary of the launch of NHK's 8K channel.

The theme of "question everything" runs through the core of Andrew Eborn's businesses and approach to life.

As Andrew points out, "Progress is enabled by questioning not why, but why not? We can now create images that look and feel like the real thing. The opportunities are tremendous across all media from retail, live concerts to totally immersive experiences."

What can now be achieved with even modest budgets is astounding. For example, Octopus TV was engaged to provide a magical window display for Hamleys. From concept to delivery Octopus TV devised and produced gloved white holographic hands which performed magic tricks mesmerising the crowds and drawing them inside to discover more.

Andrew Eborn has been involved in all aspects of the entertainment industry from content creation to distribution, managing sponsorships and licensing deals to putting on live acts at festivals and for corporate clients. Andrew has the pleasure of working with some of the biggest bands and brands on the planet. Childhood idols are now personal friends.

Andrew has helped generate revenue from every possible source including from immersive technology and even holograms immortalising several artists.

Using secrets and performance skills enhanced by Andrew's Membership in The Inner Magic Circle, he can create holograms which look, sound and smell like the real thing.

Those holograms can perform everywhere live acts can perform as well as in several places other acts can't reach. Audiences cannot tell the difference. From an artist's perspective they always give a perfect performance, looking and sounding their best. From a promoter's perspective, they are always guaranteed that the artist will show up and finish on time without the burden of a rider – the days of blue Smarties are past.

Artists can continue to earn revenue from performances long after their Fenders have faded and their Gibsons are gone. They can perform in territories never previously imagined - all on the same night!

The technology can help create perfect partnerships with more and more artists collaborating – reinvigorating forgotten stars and enabling established artists to promote and perform with the new.

The illusions can be created with existing footage which is why Octopus TV and Knot the Truth are establishing strategic alliances with producers, artists and other rights owners to open this new and exciting revenue stream.

Everyone could have the opportunity to experience Prince, Bowie, Elvis, Sinatra and Pavarotti in concert and revel in seeing The Beatles reunited.

The applications including in politics, medicine, art and the media industries are tremendous.

The future lies in embracing the exciting opportunities presented by new technology to launch brands and bands, reach new markets and open new revenue streams.

If you are interested in exploring how you could benefit from these exciting opportunities get in touch

Andrew.Eborn@OctopusTV.com

Follow Andrew on
Twitter @AndrewEborn and @OctopusTV
@KnotTheTruth

Cats
Boxing Day / Litter Box Special

Question: What do you get if you mix a classic collection of poems with a record breaking musical, a galaxy of A-list stars, an Oscar–winning director and a US$100 million budget?

Answer: A Christmas Turkey

This week saw the release of one of the most eagerly anticipated films of the year – *Cats!*

Cats is one of the longest running musicals in West End and Broadway history. It ran for 21 years in London and 18 years on Broadway.

Based on T.S. Eliot's 1939 classic poetry collection, *Old Possum's Books of Practical Cats* tells the story of a tribe of cats called the Jellicles and the night of the Jellicle Ball when Old Deuteronomy, played by Judi Dench in the movie, makes the Jellicle Choice and decides which cat will be reborn into a new Jellicle life and ascend to the Heaviside Layer.

But what is a Jellicle Cat I hear you cry?

"Jellicle cats" are first mentioned in T. S. Eliot's 1933 poem *Five-Finger Exercises*.

In Eliot's poem *The Song of the Jellicles* Jellicle cats were depicted as scruffy black and white cats. The name "Jellicle" comes from Eliot's poem "Pollicle Dogs and Jellicle Cats", where "Pollicle dogs" is an abbreviation of "poor little dogs" and "Jellicle cats" of "dear little cats."

Cats – The Mew-Vie

With a budget of more than US$100m, Oscar-winning director Tom Hooper (*The King's Speech, Les Miserables, The Danish Girl*) brought Andrew Lloyd Webber's record-breaking stage musical to the silver screen. A dream cast of A-listers with James Corden, Judi Dench, Jason Derulo, Idris Elba, Jennifer Hudson, Ian McKellen, Taylor Swift, Rebel Wilson and Royal Ballet principal dancer Francesca Hayward in her feature film debut.

The movie showcases Andrew Lloyd Webber's iconic music and a world-class cast of dancers under Tony-winning choreographer Andy Blankenbuehler.

The publicity promised that "the film reimagines the musical for a new generation with spectacular production design, state-of-the-art technology, and dance styles ranging from classical ballet to contemporary, hip-hop to jazz, street dance to tap."

What could possibly go wrong?

Everything!

Cat Fight

Midnight. Not a sound from the pavement....until the reporting embargo was lifted, coincidentally at the same time as the Trump impeachment vote hit the headlines.

A good time to release bad news.

Christmas is a wonderful time for bringing everyone together. And come together we did!

Cats was the purrfect gift to the critics.

Cats is distributed by Universal and the critics were universal in clambering for the purrfect pussy puns, hairball ho ho hos. The claws – not Santa - were out for Christmas. The best bad reviews include:

> Tim Robey, *Telegraph* - "a sinister, all-time disaster from which no one emerges unscathed."
>
> Johnny Oleksinski, *New York Post* - "a total disaster...Please wipe this movie from my Memory.'"

Peter Travers, *Rolling Stone* - "A Broadway Musical Adaptation Straight Outta the Litter Box...This disastrous attempt to bring Andrew Lloyd Webber's hit musical to the screen shouldn't happen to a dog."

Alison Willmore, *Vulture* - "To assess *Cats* as good or bad feels like the entirely wrong axis on which to see it. It is, with all affection, a monstrosity."

Peter Bradshaw, *The Guardian* - "a purr-fectly dreadful hairball of woe."

Matt Goldberg, *Collider* – "Watching *Cats* Is Like a Descent into Madness."

Jill Gutowitz, *ELLE.com* - "*Cats* Is A Nightmare That Won't End."

Kevin Fallon, *Daily Beast* – "The *Cats* Movie Is a Boring Disaster Filled With Joyless Pussies."

Richard Lawson, *Vanity Fair* - "A Tragical Mess of Mistoffelees."

John Anderson, *The Wall Street Journal* – "Going to the Dogs."

Brian Lowry, *CNN* – "*Cats* leaves behind a memory that's best forgotten."

Kevin Maher, *The Times* - "...musical mess is one for the litter tray..."

David Sexton, *The Evening Standard* - "Nearly as obscene as *The Human Centipede.*"

Twitter's Litter Tray

Seeing so many A-listers in what could be career destroying roles set social media alight. Witnessing the various actors hissing, wriggling and writhing conjured up memories of George Galloway licking a saucer of milk given to him by Rula Lenska in the UK Big Brother House's "would you like me to be the cat?"

The image of Idris Elba dressed in lycra playing Macavity the Mystery Cat caused the cinema audience to laugh out loud. Idris' tough man image was shattered with sparkling spandex.

Flaws And Claws

The digital effects are clearly unsettling creating bizarre humanoid cats. In some scenes the cats are feline-sized whilst in others they are human-sized. Most cats have been spayed and had their bumps and bulges removed apart from the brilliant Taylor Swift who is a busty Bombalurina. #Mamories

The Twittersphere cackled with titters about Jason Derulo's CGI-sized bulge. Poor Rum Tum Tugger - Tiny Tim at Christmas became Tiny Tum the todgerless Tom. The digital effects had clearly not been finished. One scene contains Judi Dench's hand with her wedding ring.

Mew – Sic

Many of the actors are not known for their singing, causing one critic to call the movie a *"tone deaf* grotesquerie."

Judi Dench was going to be the original Grizabella in the stage show before an Achilles injury forced her out. Her part was then given to Elaine Paige.

In the movie, the great Dame looked like the lion in The Wizard of Oz, coincidentally also released in 1939, the year T.S. Eliot's *Old Possum's Books of Practical Cats* was published.

But the film is not without its merits:
- Francesca Hayward, a principal dancer for the Royal Ballet, plays Victoria with beauty and grace throughout.
- Jennifer Hudson, the Oscar winning Dreamgirl, as Grizabella, the former 'glamour cat,' sings a very tearful version of 'Memories.'

- Rebel Wilson as Jennyanydots added some fun comedic moments. Rebel without a cause? Maybe. Rebel with some claws? Absolutely!
- James Corden as Bustopher Jones made me smile like a Cheshire Cat with his rendition of The Cat About Town. Corden pointed out on *The Ellen DeGeneres Show*, "I'm like Daniel Day-Lewis. I immerse the character. You know what I mean? I've just been living as a cat. I don't use the bathroom. I have a litter tray." He also confessed to Radio 2 "I haven't seen it. I've heard it's terrible."
- However, Taylor Swift confided to Vogue "I really had an amazing time with *Cats*,'" adding, "I think I loved the weirdness of it. I loved how I felt I'd never get another opportunity to be like this in my life."

How Many Lives Does *Cats* Have?

In an unprecedented move, on the film's opening day Universal advised thousands of cinemas that they would be receiving an updated version with "some improved visual effects," according to a copy of the memo obtained by *The Hollywood Reporter*. A new edit – catnip, perhaps?

Embrace The Feline Failure - Don't Cry Over Spilled Milk

The cat-astrophic reviews, however, should not put you off seeing it. Curiosity may, in fact, stop the cat being killed.

Cats may well be a dogs' dinner but rather than spay the movie we should embrace its awfulness.

Cats could turn out to be a cult classic in the same way as the 2003 independently produced *The Room* starring Tommy Wiseau.

The Room is about an amiable banker whose friends betray him one by one. It was described as "the *Citizen Kane* of bad movies" with Ignatiy Vishnevetsky of *The A.V Club* declaring it to be "the greatest bad movie of our time...*The Room* is nearly an anti-film—an inane and unintentionally surreal soap opera, filled with non sequiturs, confused characters, and gratuitous, anatomically incorrect sex."

Nevertheless, *The Room* went on to be a cult classic and also inspired the Oscar-nominated film *The Disaster Artist*.

Creaming It In

Cats may well be a Christmas turkey with lame openings at the Box Offices in the UK (£3.4m) and US (£5m) having been savaged by the critics but the long tail may yet prove it to be a success.

I predict that *Cats* will be the movie you will never fur-get no matter how hard you may try!

Predicting Failure
A Note From The Author

For several years, I have successfully predicted major events and election results with alarming accuracy.

Whilst pollsters flubbed spectacularly, I predicted the Tory victory in 2015, the backfire of Theresa May's decision to call a snap election in 2017, the twists and turns of Brexit, the unanimous decision of The Supreme Court that prorogation was "unlawful, void and of no effect and should be quashed," the post truth era of Trump, even the date of the election – 12th December 2019.

Against all the odds, my 100% record of predicting human behaviour, media coverage and results remains intact.

As a Member of The Inner Magic Circle with Gold Star, as well as a lawyer and strategist, my predictions and deep understanding of human behaviour come with the territory.

Earlier this year, I provided 7 simple steps to ensure victory in the election.

Step 1. Manage The Brand – Call In The Experts

Politics has always been a popularity contest. The key to maintaining any brand is control of the message.

Politicians are in the business of being elected.

We love to know everything about our politicians - their families, love lives, personal preferences. We snigger at their failings and love it when politicians hurl insults at each other. From the bear pit of Prime Minister's Question Time to name calling, "Mugwumps," "dadiprats" and "gollumpuseses" all.

The importance of spin doctors, strategists and brand experts has never been stronger. Having experts at your side throughout media coverage is essential.

Saatchi & Saatchi created the 1979 ad campaign which saw Mrs Thatcher become prime minister. Showing a long, snaking queue to the unemployment office with the headline 'Labour Isn't Working,' they went on to help secure three successive electoral victories for the Tories.

Where would Tony Blair have been without his Director of Communications and Strategy, Alistair Campbell?

The last election saw the return of Australian political strategist, Sir Lynton Crosby - "master of the dark political arts" - who was appointed to play a leading role running the Tory campaign. The "Wizard of Oz" was also widely regarded as the architect of David Cameron's victory.

Sir Lynton correctly predicted the outcome, in spite of the polling companies. "It wasn't just Ed Miliband's Labour Party that revealed itself as out of touch and remote from the people who are the backbone of Britain. It was a failure for the Westminster-centric 'Eddie the Expert' and 'Clarrie the Commentator' who were tested and found wanting. It was as much a judgement day for them as Ed Miliband and they lost."

Yet again, history will repeat itself and the master media manipulators will ensure victory.

Never let the truth stand in the way of a good story.

Step 2. Play The Fear Card

Fear is one of the most powerful motivators. It helped secure victory for Brexit and Donald Trump.

It has also aided the rise of the far right.

Fear for our jobs.

Fear for our position in the world post Brexit.

Fear for the environment.

The threat is clear - if you don't vote we will fail and it will be your fault!

Step 3. Adopt A Catchy, Generic Slogan

As with all advertising, a catchy slogan with sentiments which no one can dispute is key. The Labour Party's 1997 election campaign slogan "Britain Deserves Better" helped secure victory for Tony Blair. Similarly Trump's "Make America Great Again" saw The Donald defy the pollsters.

May's "strong and stable leadership" mantra was regurgitated ad nauseam when asked about everything from the scandal of food waste to Brexit to Tory tax policies.

Britain deserves better!

#StayAtHome
#StayAlert
#Cummings&Goings (Shurely shome mistake?? Don't call me Shurley etc etc)

Step 4. Limit Debate

The Nixon/Kennedy debate on 26th September 1960 elevated John F Kennedy from a relatively unknown senator from Massachusetts to a super star. In comparison Nixon seemed shifty, sickly and sweaty.

The underdog has everything to gain from a TV debate. Those with a clear lead have everything to lose.

Skilled interviewers can destroy reputations. History is littered with victims: Frost extracting an apology from Nixon for Watergate; Paxman savaging Michael Howard, then the Home Secretary. In May 1997 regarding a meeting Howard had with Derek Lewis, head of the prison service about the possible dismissal of the governor of Parkhurst Prison, John Marriott. Paxman asks Howard the same question – "Did you threaten to overrule him?" – twelve times. A masterclass in persistence resulting in a very uncomfortable stalemate.

Developing the art of avoiding the question is key and has given birth to a whole new business of self-professed experts – media trainers.

Master the art of question dodging and learn how to use lots of words but say absolutely nothing.

In his analysis of the techniques used in political interviews, Peter Bull from the Department of Psychology at the University of York, identified 35 different techniques used to evade answers from "rephrasing the question," "attacking the question," "give a non-specific answer to a specific question" and even Margaret Thatcher's favourite, "attacking the interviewer."

Step 6. Restrict The Media

"Whoever controls the media, controls the mind" — Jim Morrison

Donald Trump is undoubtedly a master of media manipulation.

It is so much easier to only allow access to those outlets who are friendly to your cause and ban others. The White House has already barred several news organisations from off camera press briefings. The BBC, CNN, the New York Times, the Guardian, the Los Angeles Times, Buzzfeed, the Daily Mail and Politico have been among those excluded.

But don't expect the media to go down without a fight.

The New York Times editorial pointed out, "That First Amendment can be inconvenient for anyone longing for power without scrutiny. Mr. Trump might want to brush up on what it means and get used to it."

The LA Times said, "If the intent was to intimidate reporters into writing fewer things that the administration does not like, and more things that it does, it is doomed to failure."

Step 7. …And If All Else Fails, Blame The Media

Follow these simple steps and election success is guaranteed.

Failing which, take The Trumpster's lead. Any view which is contrary to your own can be dismissed as "fake news" setting the stage for your spin doctors to peddle the "alternative facts."

Blame that "enemy of the people," the corrupt press, pointing out that you now realise how dishonest the media is.

Do not be surprised when you hear the announcement of TNN Trump Network News. You heard it here first folks.

Final Thought

What is shockingly clear is that at every election the biggest winner is apathy. Non-voters would play a pivotal role in the general election if they were to use their vote.

Katie Ghose, Chief executive of the Electoral Reform Society said, "Despite our flawed voting system, it's vital that everyone gets out there and uses their hard-fought right to vote. Voting does make a difference, and wherever people are they should have their say. The point is to make the voice of voters even stronger."

We need to restore trust in politicians and to bother about the issues – climate change, food waste, housing, caring for an increasingly ageing population, health and education for all, security as well as our position on the global stage after Brexit.

Focusing on the real issues whilst being aware of the tactics being deployed to manipulate the media is a start.

Whatever your views, we should encourage everyone to question everything and to vote.

Google Glass

ELECTION SPECIAL
SEEING RED OR TURNING THE AIR BLUE!
WILL FRIDAY 13TH BE A NIGHTMARE ON DOWNING STREET?
FIRST PUBLISHED 12TH DECEMBER 2019

Today's the day we go to the polls in what has been dubbed The Brexit election, The Great British Break Off.

Predictions

With a nation crippled by Brexhaustion, will it be brake off time? Will Britain come out of neutral and get Brexit done?

Well Hung?

This has been described as "arguably the least predictable election in modern history."

Chris Curtis, YouGov's political research manager, points out, "The margins are extremely tight and small swings in a small number of seats mean we can't currently rule out a hung parliament. As things currently stand, there are 85 seats with a margin of error of five per cent or less."

I predict that there will be widespread tactical voting. And sure enough, this week saw Anthony Joshua (AJ) retain his heavy weight title. As also previously pointed out throughout this rollercoaster of an election, I predicted that, like AJ, BJ (Boris Johnson) will retain his heavy weight belt.

Don't be surprised to see the largest Tory majority since the Eighties, beating John Major's 21-seat margin of Commons control in 1992.

I also predicted Labour's worst result for more than 30 years.

I predicted some upsets, but a 7% margin give or take the usual 2% discrepancy is viable...you read it here first, folks!

What I also predicted was that this election would again shine a light on the increasing power of social media platforms being used as the vehicles for everything from Google Jacking to the cunning Conservatives being accused of misleading the public by renaming the party's official Twitter account 'factcheckUK' to fake news stories such as a claim that Lib Dem leader, Jo Swinson, enjoys murdering squirrels with a catapult.

Nuts!

Those global gorillas Google, Facebook, Twitter *et al* play a huge role in influencing voters.

As I previously pointed out, polls are often apart from reality.

Perception is the ultimate deception.

I agree with the journalist, Peter Hitchens that, "Opinion polls are a device for influencing public opinion, not a device for measuring it. Crack that, and it all makes sense."

As I said, the threat of a hung parliament or a perceived opportunity to win in marginal seats spurs people on to vote.

Talk of landslide victories only serves to breed complacency. Why bother to vote if it's not going to make a difference?

That is the reason you will see fear being used to ensure that we get out and vote.

So why do the polls get it wrong so often?

I suggest that the polls get it wrong where people don't want to admit publicly to what might be perceived as an unpopular view for fear of repercussions. This would account for the fact that so many people dismissed the possibility of Trump.

Against all of the odds, I predicted the rise of Teflon Trump.

I also predicted that he would survive impeachment.

Borat's Faceoff With Facebook

The Wild West Web provides a perfect propaganda platform as it's unregulated and unchecked.

The International Committee on Disinformation and Fake News was told that the business model adopted by social networks made "manipulation profitable."

And indeed it is. Political ads are a fantastic money spinner for the social media giants.

Ali G & Borat star Sacha Baron Cohen, addressing The Anti-Defamation League's Never is Now summit in New York, criticised Facebook boss Mark Zuckerberg who in October defended his Facebook's position not to ban political adverts that contain false or misleading information.

"If you pay them, Facebook will run any 'political' ad you want, even if it's a lie. And they'll even help you microtarget those lies to their users for maximum effect." Sacha emphasised, "Under this twisted logic, if Facebook were around in the 1930s, it would have allowed Hitler to post 30-second ads on his 'solution' to the 'Jewish problem'."

The Ali G star also criticised Google, Twitter and YouTube for pushing "absurdities to billions of people." He went on to say it was time "for a fundamental rethink of social media and how it spreads hate, conspiracies and lies."

Some social media giants have taken steps to address concerns. Twitter announced in late October that it would ban all political advertising globally from 22 November 2019.

Google said it would not allow political advertisers to target voters using "microtargeting" based on browsing data or other factors.

The term "microtargeting" was coined in 2002 by political consultant Alexander P. Gage. Microtargeting is where direct marketing datamining techniques are used that involve predictive market segmentation to track individual voters and identify potential supporters. That is why you saw so many targeted ads every time you checked Facebook.

Google is proud of the steps it takes to support democratic processes around the world.

The fact remains that the power of social media giants will only intensify – ignore at your peril!

But not everything the social media Midas giants touch turns to gold as we see with this week's nomination for The Octopus TV Failure Awards, Google Glass.

Google Stats

Founded in September 1998, Google has secured its position as a global gorilla and is busy acquiring more and more of the jungle battling the Competition Commission along the way.

The statistics are both impressive and terrifying:

Google's parent company Alphabet saw revenue for the twelve months ending September 30, 2019 rise to $155.058B, a 19.4% increase year-over-year.

Google does not publish its search volume data. It is, however, estimated that Google processes approximately 70,000 search queries every second, the equivalent of 5.8 billion searches per day and approximately 2 trillion global searches per year. The average person– whomever that is - conducts between three and four searches each day.

Google now enjoys almost 93% market share for search.

The landscape of the jungle in which Google is the king is forever evolving.

The Future Today

I am often invited to speak around the world on what the future holds in terms of technological developments and to predict trends and human behaviour helping businesses not only to survive but to flourish.

I have just returned from Cannes where I was discussing The Octopus TV Failure Awards TV series and Octopus TV's / Knot the Truth's rich slate of diverse programmes across each genre.

In Cannes I was engaged by Japanese broadcaster NHK to help present their 8K content to the world.

8K is as much as the human eye can cope with. The next stage which we are exploring will be to feed directly into the brain.

As I predicted, the inevitable march of AI and robotic technology is changing the way we live and work replacing millions of jobs.

Jack Ma, chairman of Alibaba, has predicted decades of "pain" due to new technologies.

Elon Musk, the Tesla founder, was in the news this week again having won the US$190m defamation suit against him over his Tweet saying that humans need to merge with machines if we want to stay relevant.

Wearable Technology Market

Wearable technology is the first step to our cyborg transformation.

Analyst Gartner predicted that in 2020 consumers worldwide will spend $51.5 billion on wearable tech, including smartwatches, ear-worn devices, and headsets, a 27% jump from 2019's forecast spending. [Note July 2020: That prediction was BC ie Before Corona. Hindsight is not foresight.]

Through The Looking Glass

Google's deep pockets in the second decade of the new century put it in the perfect position to assist our transformation to cyborgs. Google can afford to fail.

It was inevitable that Google would look at products to help feed the machine whilst turning us into one, all the while continuing to gather more and more information about us.

If content is King, data is its Queen and Google the royal carriage for carrying content and delivering data.

The Google X lab was set up to devise and bring to fruition futuristic ideas including indoor GPS, the Google Brain and wearable technologies. At 1489 Charleston Avenue, the lab's first project was born: later known as Google Glass the computer you wear on your face with built in camera, display and microphone. The head-mounted display and voice activation capability enable users to navigate the internet in the same way as a hands-free smart phone. It started, according to Google, as "little more than a scuba mask attached to a laptop." Rather than perfecting the device and ensuring it worked before general release, Google co-founder, Sergey Brin, wanted to release Google Glass as soon as possible arguing that consumer feedback would improve the design. A strategy that coupled with the right PR and marketing spin could work well, making consumers feel they were part of the development family and accordingly happy to fork out their well-earned cash on the never-ending release after release after release.

The release of the beta Google Glass was not a quiet affair. As The New York Times pointed out, "Google Glass didn't just trickle out into the world. Instead, it exploded with the kind of fuss and pageantry usually reserved for an Apple iSomething."

In July 2012 Sergey Brin unveiled Google Glass with a group of skydivers jumping from a zeppelin above San

Francisco. Brin was applauded as a real-life Tony Stark from Iron Man.

To reinforce the fact that Google Glass was a work in progress, the first version was not made available in retail stores but rather limited to 8,000 "Glass Explorers" who paid $1,500 for the privilege of being an early adopter.

Early in 2013 those interested in being a Google Explorer were invited to tweet with the hashtag #IfIHadGlass. It would be great to track down those who used the other Octopus TV Failure Awards nominee Twitter Peek to do so. Let me know.

The media burst a blood vessel with excitement.

Time Magazine named Google Glass one of the "Best Inventions of the Year."

Fashion went four-eyed when Diane von Furstenberg featured Google Glass in her 2012 New York Fashion Show.

Vogue magazine gave Google Glass its own 12-page spread.

Everyone rushed to provide their own reviews if only to boast of having been able to get their hands on a pair.

Further exposure was guaranteed by having numerous celebrities and other public figures don Google Glasses including Lady Gaga, Samuel L. Jackson and Donald Trump's favourite, Meryl Streep. Even Presidents, Princes and Princesses were pictured with a pair from President Obama to Princes Charles & William.

Specs And The City

Glass' potential use in journalism was also explored. *Voice of America* Television Correspondent Carolyn Presutti and VOA Electronics Engineer Jose Vega launched "VOA & Google Glass," and The University of Southern California even offered a course on "Glass Journalism."

Shattered Dreams

Blinded by the hype and hysteria, it was easy to forget that Google Glass was still in open beta. The attention generated was massive and expectations were high. It was therefore inevitable that, as the product was not yet fully finished, that there would be a backlash.

From being the most sought after gadget in the whole of geekdom, Google Glass became the target of ridicule.

Google Glass provided perfect comedic fodder and was featured in several shows from Saturday Night Live to "Oogle Goggles" in "The Simpsons."

Reviewers were not backwards in their condemnation. Google Glass was described as "the worst product of all time," "plagued by bugs," "of questionable use," "overpriced" and having an abysmal battery life.

People also complained of being disoriented and having headaches.

Google Glass went from being the must-have product to being "socially awkward."

The brand also suffered as users started to be referred to as "Glassholes."

Privacy concerns were raised, with people afraid of being recorded during private moments. A number of establishments banned users wearing Google Glass. No Glassholes!

There were also concerns over potential eye pain caused by users new to Google Glass including some comments reported from Google's optometry advisor Dr. Eli Peli of Harvard, although predictably there were also reports of backtracking on those early comments.

I am working with Ananth "Vis" Viswanathan, Consultant surgeon at Moorfields Eye Hospital on The Octopus TV Health series. Vis warned of the possibility of Google

Glass distracting the user which could be "potentially dangerous, for example whilst driving."

Broken Glass

On 15 May 2015 Google announced that it was closing the Glass Explorer programme.

Several believed that this was the end of Google Glass. In its obituary for Google Glass, one reviewer wrote "Google Glass promised many things, but in its brief and over-hyped lifespan achieved only one thing of note: the creation of the word "Glasshole."

If necessity is the mother of invention, Failure® is the father of success.

In its announcement of the closure of Glass Explorer, Google hinted that it was going to "focus on what's coming next…"

….and so more than 2 years after the end of the consumer version Alphabet, Google's parent company, announced in July 2017 that it was re-introducing Google Glass with more practical industrial applications and now simply calling it "Glass."

Its website proudly proclaims: Glass is a hands-free device, for hands-on workers and can be used by everyone from doctors to automobile assemblers. Glass intuitively fits into your workflow and helps you remain engaged and focused on high value work by removing distractions. A quick 'OK Glass' can [enable users to] access training videos, images annotated with instructions, or quality assurance checklists that help you get the job done, safely, quickly and to a higher standard. Glass can connect you with co-workers in an instant, bringing expertise to right where you are. Invite others to 'see what you see' through a live video stream so you can collaborate and troubleshoot in real-time."

John Naughton pointed out, "The rebirth of Google Glass shows the merit of failure."

The final nail in the coffin of the consumer version of Glass came this week, however, when it was announced that Google plans to put out one final software update and then all those Explorers will be left to fend for themselves.

"Glass Explorer Edition is receiving a final update that you will need to manually install. After February 25, 2020, this update removes the need and ability to use your Google account on Glass. It also removes Glass' connection to backend services."

The update simply lets you pair Glass with the phone. MyGlass will stop working. Bluetooth will continue, as will the ability to creepily take photos and videos via your lenses. Those Glassholes who refuse to apply the update will still be able to use it but mirrored apps such as Gmail, YouTube and Hangouts will not work.

Google may have pulled its consumer version of the AR glasses but for now it is nevertheless still experimenting with enterprise versions for businesses to use in industries including logistics and manufacturing.
The Enterprise Edition of Google Glass is still alive.

Of note is that Glass paved the way for others wanting to make passes at the business of AR glasses.
We saw the launch of the much cooler Snap's Spectacles - Snap's first hardware product which retailed initially at only US$129, less than 1/10 of the cost of Google Glass.

2019 also saw several global giants enter through the AR looking glass.

At the gloriously named Snapdragon Tech Summit in Maui, Qualcomm announced it would be teaming up with Pokémon Go developers Niantic to make AR reference hardware and software for augmented reality glasses.

Apple is looking at a 2023 release date for a pair of augmented reality (AR) glasses set to follow the release of an AR headset in 2022.

Facebook is also set to expand its inventory with augmented reality glasses, Facebook's Facewear.

2019 also saw Amazon unveil Echo Frames, their Alexa-powered set of smart glasses.

At CES earlier in 2019, Vuzix released its Blade AR smart glasses, which were described as "the next-gen Google Glass we've all been waiting for." The glasses feature an 8-megapixel camera, Amazon Alexa for hands-free use, and prescription inserts so you can wear them without glasses. The Blade glasses recently picked up the CES 2019 Innovation award for outstanding design and engineering

Fashion And Function Frames

Social acceptance of AR glasses is a barrier.

To assist companies to succeed they should focus on fashion as well as functionality.

Amazon-backed start-up North, for example, has created a more fashion-focused version of the tech-enabled specs.

Whilst the success of the individual products is uncertain, the space itself is only going to grow as more companies experiment with the possibilities.

International Data Corporation (IDC), predicted before Corona (BC) that worldwide revenues for the augmented reality and virtual reality (AR/VR) market would grow from $5.2 billion in 2016 to more than $162 billion in 2020.

Once social acceptance increases, including by fusing function with fashion, I predict that the social media platforms' share of that market is likely to be eye watering.

The early introduction of Google Glass in its open beta form was, however, a failure and is therefore this week's nomination for The Octopus TV Failure Awards.

In the meantime, as we enter a new chapter in this most delicious and bizarre era of Politics make sure we don't let others view the world through rose tinted glasses, but rather put everything in focus next year with 2020 AR vision!

NOTE

My prediction prior to the election received extensive international media coverage.

My 100% record of predicting human behaviour, the media coverage and results remains intact!

YouGov's political research manager Chris Curtis described the election as "arguably the least predictable election in modern history."

Against all the odds, I successfully predicted that the Tories would secure the largest majority since the days of Margaret Thatcher in the 80's.

Harley-Davidson Eau de Toilette
"Hog Wash"

This week in our series on nominees for The Octopus TV Failure® Awards (TOFA) we look at a product from an iconic brand no doubt made dizzy by the sweet smell of success and hoping to bottle that scent - Legendary Harley-Davidson eau de toilette.

Harley History

Harley-Davidson is the embodiment of the American dream. Harley's history is perfect movie fodder... (watch this space).

In 1901 a 21 year old William S. Harley completed a blueprint drawing of an engine designed for use with a regular pedal bicycle. Over the next 2 years Harley continued work on this "motor bike" together with his childhood friend Arthur Davidson and his brother Walter Davidson. They used the north side Milwaukee machine shop/shed at the home of their friend, Henry Melk. The first Harley-Davidson® motorcycle was finished in 1903.

That first motorcycle was, however, not powerful enough to climb the hills around Milwaukee, Wisconsin without additional pedal assistance. Work therefore continued on producing a more powerful engine and loop-frame design moving from the motorised bicycle lane to the superhighway of future motorcycle designs.

Harley-Davidson incorporated in 1907. In that year William Davidson quit his job with the Milwaukee Road Railroad to join forces with his two brothers and Harley. Harley-Davidson began hiring employees and by the end of the decade was producing bikes using its signature 45-degree air-cooled V-Twin.

Racing – Without Pulling A Ham String

In 1914 Harley-Davidson started motor racing enjoying considerable success. In 1922 Harley-Davidson riders went on to sweep all eight National Championship Races.

HOG

The "hog" association started in 1920 when a team of farm boys, including Ray Weishaar, used to do a victory lap with the racing team's mascot, a pig, after each race won by the team.

On August 15, 2006, Harley-Davidson Inc. had its NYSE ticker symbol changed from HDI to HOG.

Harley Today

Today Harley-Davidson employs more than 6,000 people and generates sales of US$6 billion. According to Forbes Harley was ranked #97 in Best employers for women 2019. Its corporate headquarters are located near the site of the famous shed where William S. Harley and the Davidson brothers created the first motor bike, just west of downtown Milwaukee.

Living Pieces Of American History

Harley-Davidson are proud of their motorcycles and the loyalty and passion generated – and rightly so:

"For us and for our loyal customers, the motorcycles we build aren't just motorcycles. They are living pieces of American history, mystique on two wheels. They are the vehicle with which our riders discover the power, the passion, and the people that define the Harley-Davidson Experience.

To make America great again you would ride in on a Harley.

Knucklehead, Panhead, Shovelhead And Evolution®

"From the very earliest days, Harley-Davidson has favoured evolution of its powerful engines over radical change. The Knucklehead, Panhead, Shovelhead and

Evolution® engines are all deeply immersed in the heritage of the Motor Company and the hearts of our enthusiasts. The Twin Cam 96® and Twin Cam 96B™ engines are now securing their own place in company folklore while the powerful Revolution® engine in the stunning V-Rod® motorcycle is taking Harley-Davidson down roads never travelled before."

Harley Owners Group®

In addition to providing support to their dealers through national advertising campaigns and promotions, Harley-Davidson offer numerous programmes designed to drive customers into their dealerships.

The Harley Owners Group, or H.O.G.®, (Gedditt? Got It? Good!) is the largest manufacturer sponsored motorcycle club in the world and it continues to grow.

Events, Rental & Tour Experiences

To provide additional revenue streams as well as promotion for the brand, Harley-Davidson put on various events and have an impressive rentals and tours' programme recognising that "a positive rental or tour experience is a logical step toward ownership."

Brand Mythology & Lovemark Loyalty

As Kevin Roberts, Saatchi and Saatchi's former worldwide chief executive officer and Chairman pointed out the most powerful brands are those that have built their own mythology. Truly successful brands don't have 'trademarks' they have 'lovemarks.'

Harley-Davidson was cited by Kevin Roberts, as a perfect example of a "lovemark" in Fast Company magazine's September 2000 issue.

Branded With The Brand

Harley-Davidson customers are undoubtedly passionate about the brand. Some proudly sport tattoos of the Harley-Davidson logo. A HOG tattoo is so much more powerful than a Peppa Pig tattoo for all you swinestone cowboys out there.

Motorcycle Merchandise Madness

Like many brands, Harley-Davidson offer a range of products with a smaller price tag than their bikes. This helps drive footfall to the dealerships as well as provide additional revenue streams and keeps the aspiration of a Harley ownership alive.

Harley-Davidson have been selling clothing designed for both on-road and off-road use since 1912. Dealers now have a variety of items "created to appeal to a variety of tastes" and bring both riders and non-riders to dealerships to "greatly enhance… profitability."

Today, from Harley-Davidson® MotorClothes® to licensed merchandise, dealers are provided with a variety of items from which to choose. From an extensive line of stylish Harley-Davidson® and Buell® clothing designed by motorcyclists to a broad range of Harley-Davidson® and Buell collectibles.

True Lovemark

Want to get married overlooking a showroom full of America's favourite motorcycles?

Everything is possible!

The Not-So-Sweet Smell Of Success

Brand loyalty is the Holy Grail for businesses. Testing that loyalty can, however, be dangerous.

In 1994 Legendary Harley-Davidson eau de toilette was launched under licence to Frangrantica.

This was a brand extension too far. Many customers who loved Harley-Davidson hated the Disneyfication of the brand.

Joe Hice, former director of corporate communications for Harley-Davidson, acknowledged the failure: "Over the years we've tried a number of different approaches to merchandising and have put the Harley-Davidson brand on some things that in retrospect we may not have been well-advised to do. The company is much more selective today about how we go about extending the brand."

Why Did It Fail?

Brand extensions can be very successful, enabling new products to be introduced under a well-established existing brand name.

As I pointed out, when looking at Colgate Lasagne, generally to benefit from the existing brand the new products need to be consistent with core brand values and customer perceptions and expectations. What values and expectations are conjured up when customers think of Colgate? Colgate is strongly associated with health and oral hygiene. The very name and logo instantly suggest that fresh, minty taste. Associating the brand with food was clearly not going to work. Who wants toothpaste flavoured pasta?

Hog Wash

So what values and expectations are conjured up when customers think of Harley-Davidson?

As Charles" Chuck" Brymer, former chief executive officer of the Interbrand Group and now President of DDB Worldwide points out, "Harley-Davidson values are strong, masculine, very rugged...for Harley-Davidson to go into a sector that doesn't live up to what those values are would be disastrous."

A Harley-Davidson perfume for all those who want to smell like bikers, hot and sweaty from getting their kicks on Route 66!?!?

Even a French voice over in the ad purring about 'arley Davidson failed to destroy the sensory association – Do you really want to smell like a Harley?

What Next?

A Harley-Davidson shampoo with that fresh petrol aroma for really oily hair? Harley-Davidson Toothpaste for that snoggable HOG's breath? Hmmmm.

Lessons To Be Learned

I have had the pleasure of working with the strategic development of brands for several years including assisting with the generation of additional revenue streams.

Properly executed, brand extensions undoubtedly work. It is essential, however, that businesses understand the true nature of their brand.

Line extensions inconsistent with brand values can be disastrous and also adversely affect other core products.

Jurassic Pork

Don't alienate your core customers. As with reputation and trust, loyalty and love bands take years to build but can be destroyed in seconds.

Disgruntled

Testing loyalty by seeking to squeeze every dime from fans through inappropriate licensed products may not only result in brand dilution but also loss of customers - the lifeblood of your brand.

Mistaken Bacon

Less is often so much more. Slapping your brand on as many products as possible is a mistake and merely serves to devalue your brand.

Hog Roast

The very name Harley-Davidson conjures up strong feelings and associations. The idea of Harley-Davidson perfume, however, stinks. It's the pits!

Pigs might fly before the HOG perfume would ever take off. It's had its bacon.

Market research would make it clear that this is a brand extension way too far. It would not take Ein-swine to figure that out.

There's no sizzle in that sausage. It's HOG Wash!

Trump - The Game

PUBLISHED NOVEMBER 2019

With the impeachment inquiry underway this week we look at a product featuring the POTUS guaranteed to gain more column inches in history than any other: Trump – The Game

Where Would We Be Without Donald Trump?

Contrary to all the pollsters, I successfully predicted that The Trumpster would make it to The White House winning me money from the head of a major media company in the process – it was news to Fox.

Do not be surprised when he secures a second term.

Every day The Donald never fails to deliver more and more gold for journalists, brands, comedians and creatives.

Trump is the master media manipulator.

Trumpeting Trump

Like him or loathe him, one thing for sure is that Trump is a powerful brand.

The market is flooded with Trump merchandise both licensed and unlicensed.

From the 12 inch Donald doll from Stevenson which utters 17 Trump recorded phrases including such classics as, "I should fire myself just for having you around."

To the usual caps, mugs, T-Shirts, badges and other merchandise on the campaign trail to Making America Great Again.

All the way to Trump scented candles.

Given how much money can be generated from the lucrative board games market, a Trump board game was inevitable.

Board? This'll Wake You Up...

Trade analysis magazine *ICv2* estimated that in 2015 the card and board games market was worth US$1.2 billion in US and Canada alone.

The global board games market is predicted to be worth almost US$5 billion by 2021.

Impressively, in 2015 over US$200m was raised though crowd funding platforms for board games. On Kickstarter, tabletop games made 6 times more than video games in the first half of 2016. The game Exploding

Kittens raised almost US$9milion making it one of the most successful projects since Kickstarter was founded in 2009 together with a reprint of another board game, Kingdom Death Monster, which raised more than $12m.

Smorgasbord Of Services

It is a business I know well. I am involved with several companies across the IP value chain from idea creation, production, distribution and new technology including holograms, offering a smorgasbord of services. As part of that value chain, I am Head of Commercial & Strategy of Boxatricks with the creators of some of the biggest game shows in history including Who Wants To Be A Millionaire and The Weakest Link. We have several new and exciting formats and are kept busy devising the next big thing. In addition, we are engaged as format doctors to help others maximise the return on their rights across all media including bringing their formats to the screen as well as creating shows for brands and existing IPs such as turning the card game Top Trumps into a TV game show. "Get Carter?" "Top Trump!"

Shelf Life

Board games have a long life cycle - an average of 8 years. Some last for generations. Monopoly, for example, was launched in 1935 and is still one of the most popular board games enjoying its highest number of sales more than 80 years after it was introduced.

The revenue that can be generated from the strategic licensing of intellectual property is significant. Get it right and the world is your lobster.

Trump: The Game

With this in mind, Jeffrey Breslow, a leading games inventor and former President and CEO of Big Monster Toys, had the idea for a board game based on everyone's favourite Big Monster, Donald Trump, and his 1987 book "The Art of The Deal."

Breslow pitched the idea to Trump and suggested splitting the profits equally. Trump replied, "I don't do 50-50." It was agreed that Trump would receive 60 to Breslow's 40 percent. Breslow said, "The game wasn't sellable without Donald Trump. He could have squeezed me for even 80-20. He knew he was in the driver's seat."

It will upset The Donald to read that he could have got more.

The Game was launched by Milton Bradley Company at Trump Tower on 7th February 1989. At the launch The Donald announced that he was not looking to make money from the game himself rather his percentage of the game's profits would be donated to charity. "The game was just ego to him, one more promotion," Breslow said.

In the predictable fear and smear in the last few days of the general election campaign, during the cut and thrust of the US Presidential election campaign The *Huffington Post* had the audacity to question Trump's claimed charitable donations.

Trump Has A New Game

"It's not whether you win or lose, it's whether you win!"

Trump was happy to help promote the game and even appeared in the TV commercial.

Trump – The Game is a roll and move game about buying and selling real estate. It was described as a boring and complicated variation of Monopoly although Trump – never backwards in coming forwards - pointed out, "I really like the game. It's much more sophisticated than Monopoly."

Trump: The Game Flopped Shifting Just 40% Of Its Expected 2 Million Sales.

Milton Bradley President George Ditomassi said, "The game was just nailed to the shelf." Trump himself admitted the game was "too complicated."

In 2004, following the success of The Apprentice, however, Parker Brothers re-released the game with simplified rules and a new tagline: "It takes brains to make millions. It takes Trump to make billions."

Players make their way round the board with plastic "T" shaped markers buying and selling real estate trying to make more money than the other players. Each player begins with US$500 million. The smallest denomination is US$10m.

The object: Bluff opponents into spending foolishly while you buy low and sweep up big profits.

As a signed letter from Trump states, "The object of the game is to make hundreds of millions of dollars. If you are clever, aggressive and lucky, you could end up with a billion or more!"

Trump is everywhere in the game. His face is featured on the money and the game's most important cards are the "THE DONALD" and "YOU'RE FIRED."

Trumpian truisms appear on each "Trump card" such as "I would fire the person most likely to fire me."

Game Over

Breslow recognised that Trump: The Game - whilst tipping its hat to Monopoly - was never going to enjoy the same success. He told the Washington Post, "A huge percentage of those games were never taken out of the box. It was bought as a gift item, a novelty, a curiosity. Trump got that. He had zero interest in how the game played."

...and that's about the level of interest most of those who did play it managed to muster...

Whilst the *Chicago Tribune* felt "It's a sophisticated acquire-and-deal game" others were not so kind. The *Los Angeles Times* said it was: "not the kind of thing you want to pull out on the spur of the moment when grandma comes over."

Another reviewer said "I loathed every miserable second of it."

Phil Orbanes, senior vice president of research and development for Parker Brothers, pointed out that Trump's game, "...can leave you exhausted and feeling like you don't want to play again. As accurate as it may be at capturing the feeling of insecurity in the real world, the game doesn't give you a feel-good experience, which is the purpose most people rely on for playing games."

Time magazine listed the game among the Donald Trump failures and *Fortune* included the game in its list of five top Trump business failures."

"A Great Game If You Don't Have Very Many Friends"

Mother Jones magazine summed up the feeling of several critics "This is a great game if you don't have very many friends...the game's flaws—its erratic nature, its contradictions, its singular obsession with the rapid accumulation of wealth for the purpose of acquiring luxury real estate and firing people—are also Trump's flaws." Trump "...basically took Monopoly money, stuck his face on it, and added a bunch of zeroes."

"Teflon Trump: The President's Game Avoiding Impeachment" Cunningly Devised By Andrew Eborn

It has been 15 years since the last re-launch...and a lot has happened since. It would be interesting, therefore, to update the product to "Trump – The President's Game" Third time lucky perhaps ……

In your quest to Make America Great again, the cards saying whether you move backwards or forwards in the game could include:

- Benefit from hacking by Russia and/or Kiev
- Caught engaging in locker room banter
- Caught pussy grabbing
- Hurl new insults at "crooked Hillary" and anyone else who dares run against you.
- Claim the largest crowd ever in the history of crowds

- Roll, move and auction your way to Mexico paying for a wall
- Sack your staff – when your spin doctors are the story you know you're in trouble. Scaramucci, Scaramouche, will you do the Fandango?
- Claim to invent "fake news"
- Blame the fake news media for any story you don't like
- Sack your staff, again
- Tweet throughout
- Invent a new language – cup of covfefe, anyone?
- Call multi-Oscar winner, Meryl Streep "overrated"
- Hold Theresa May's hand to stop her running through corn fields creating more crop circles
- Tell Macron's wife she is in "great shape"
- Threaten to tear up the deals on nuclear weapons / climate change etc, etc
- Add more hot air to the baby balloon
- Break Royal protocol and walk in front of the Queen
- Continue to test Constitutional customs
- Re-write The Constitution of the United States to ensure that there is no limit to the Presidential term
- Consume your body weight in Big Macs
- Set up Trump Network News (TNN)
- Establish transactional relationships with anyone who can advance the Trump brand

- Seek political favours from a foreign power "History and experience prove that foreign influence is one of the most baneful foes of republican government," wrote the first President, George Washington, in his farewell address.
- Call the Ukrainian President
- Encourage Ukraine to investigate former Vice President Joe Biden and his son, Hunter
- Go talk to Rudy
- Get over it
- Don't let the truth stand in the way of a good story
- Flood the media with a trove of conspiracy theories, distractions and misinformation designed to confuse the public and fog clarity
- Dismiss investigation as a biased, politically motivated hoax and a witch hunt
- Recognise offence is the best form of defence
- Recognise perception is deception © Andrew Eborn
- Avoid Impeachment
- Pick the winner of The British General Election

The Game would be won when you get re-elected or the world ends - whichever is the sooner.

Lady Doritos
LEAP YEAR SPECIAL
PRIDE AND PREJUDICE

Indra Nooyi, the former C.E.O. of PepsiCo, is a great role model.

Nooyi was born in Madras and went to college and business school in India. She played cricket and lead guitar in an all-girl rock band.

In 1978 she was admitted to Yale School of Management and moved to the US.

Nooyi joined PepsiCo in 1994. In 2006 she became President and CEO, and in 2007 she became Chair.

For many years Nooyi has been listed in the 100 most powerful women in the world.

Results under her tenure as CEO of PepsiCo speak for themselves. PepsiCo's revenue grew from $35 billion in 2006 to $63.5 billion in 2017.

Until the early 1990s, PepsiCo was mostly about selling soda and salty snacks. PepsiCo then began to diversify. PepsiCo went on to group its products into three categories:

"Good for You" products include fruits, vegetables, whole grains, nuts with an emphasis on less sugar and fat such as Quaker Oats and Tropicana;

"Better for You" includes lower-calorie sodas and baked potato chips such as Baked Lays potato chips, Stacy's Pita Chips, and Smartfood popcorn; and

"Fun for You" includes Pepsi, Mountain Dew, Lays and Cheetos.

In an interview with *Freakonomics Radio* in 2018 Nooyi pointed out, "... consumer tastes change all the time. They change in food and beverage faster than they change anywhere else."

Nooyi went on to say that PepsiCo, "...looked at consumer trends and we looked at where we thought the markets were growing, we knew we had to retool our portfolio. That was just not even a question. We knew that if we didn't do it, our future was in jeopardy."

Embracing Change

Nooyi witnessed the resistance to change. "When people say culture eats strategy, I lived it first-hand, because I saw how many people sort of said, 'Why should we change our company that's been so successful for a future we don't quite understand?'"

Nooyi pointed out that there is a difference in the way men and women eat their snacks.

"When you eat out of a flex bag — one of our single-serve bags — especially as you watch a lot of the young guys eat the chips, they love their Doritos, and they lick their fingers with great glee, and when they reach the bottom of the bag they pour the little broken pieces into their mouth, because they don't want to lose that taste of the flavour, and the broken chips in the bottom."

When It Comes To The Crunch

Nooyi suggested that, "…women would love to do the same, but they don't. They don't like to crunch too loudly in public. And they don't lick their fingers generously and they don't like to pour the little broken pieces and the flavour into their mouth."

Nooyi went on to point out that PepsiCo was looking at snacks that could be designed and packaged differently for women and said, "We're getting ready to launch a bunch of them soon. For women, low-crunch, the full taste profile, not have so much of the flavour stick on the fingers, and how can you put it in a purse? Because women love to carry a snack in their purse."

The Sun reported Lady Doritos as fact rather than speculation. Don't let the truth get in the way of a good story!

Predictably, media hysteria set in and the Twittersphere melted.

Little did we know that all this time we've been eating man chips.

As Ellen DeGeneres pointed out, "Who does not want a crunchy chip? If it's not crunchy it's a wet potato. After Pepsi make a chip that's not crunchy they will make a soda that's not wet."

"What Do We Want? Quiet Chips!"

Ellen went on to point out, "Women were not wronged by Doritos they were wronged by a giant orange Cheeto."

Soon, PepsiCo was insisting it was all a misunderstanding. "The reporting on a specific Doritos product for female consumers is inaccurate. We already have Doritos for women — they're called Doritos, and they're enjoyed by millions of people every day."

Lady Doritos never saw the light of day although other gender specific products have been released with a predictable backlash and have accordingly also earned nominations for The Octopus TV Failure® Awards. Watch this space.

Leap Of Faith

This Saturday will be the 29th February.

29th February is the traditional day when women can propose to men. This dates back to 5th century Ireland and 'St Bridget's Complaint.' St Bridget complained to St Patrick that women had to wait too long for a man to propose. It was agreed that the role could be reversed on 29th February.

On 29th February 1288 Scotland allowed women to propose to the man of their choice. Tradition had it that any man who declined a proposal in a leap year must pay a fine which could range from a kiss to payment for a silk dress or a pair of gloves.

In this age of equality / #MeToo the media backlash to this anachronistic tradition is predictable... watch those news reports on Saturday!

Re: Lenting

It's Lent when traditionally we give up something. This Leap Year Lent let's give up prejudice and celebrate our diversity.

We should celebrate our diversity – vive la difference - without prejudicing against it.

And so, Ladies & Gentleman, non-binaries, and everyone else I have missed, break open those Lady Doritos...just remember to crunch quietly.

Colgate Lasagne

Colgate-Palmolive, the small soap and candle business that William Colgate began in New York City early in the 19th century is now - more than 200 years later - a truly global company serving hundreds of millions of consumers worldwide.

Ian Cook, chairman, president, chief executive, chief cook and bottle washer points out, "Our 200-year history reflects the strength and innovation that our people have used to constantly transform our company and identify new opportunities."

Colgate is very proud of its rich 200 year history and rightly so.

On its website, Colgate highlights several key events over the last two centuries including:

> **1817** First Colgate advertisement appears in a New York newspaper.
> **1896** Colgate introduces toothpaste in a collapsible tube.
> **1947** Ajax cleanser is launched, establishing a powerful now-global brand equity for cleaning products.
> **1966** Palmolive dishwashing liquid is introduced. Today it is sold in over 35 countries.

1968 Colgate toothpaste adds MFP Fluoride, clinically proven to reduce cavities.
1983 Colgate Plus toothbrush is introduced. Today over 1.6bn Colgate toothbrushes are sold annually worldwide.

There is, however, no mention in this impressive list of the launch in 1982 of Colgate Kitchen Entrees, a range of frozen ready meals.

According to a number of sources, Colgate wanted to enter the lucrative ready meal market harnessing the strong brand loyalty it had so successfully developed – the hope apparently being that customers would enjoy eating Colgate dinners and then brushing their teeth with Colgate toothpaste.

Why Did It Fail?

Brand extensions can be very successful, enabling new products to be introduced under a well-established existing brand name. Apple is a prime example at the top of the tree.

Generally, to benefit from the existing brand, however, the new products need to be consistent with core brand values and customer perceptions and expectations. What values and expectations are conjured up when customers think of Colgate? Colgate is strongly associated with health and oral hygiene. The very name and logo instantly suggest that fresh, minty taste.

Associating the brand with food was clearly not going to work. Who wants toothpaste flavoured pasta?

What Could Have Been Done Differently?

Generally, products that have nothing in common should not be linked. Colgate should have launched with a different brand name in the same way they maintain Hills Pet Nutrition as a separate brand having acquired the company in 1976.
Market research would have made this clear.

Fake News?

Colgate Kitchen Entrees do not feature in Colgate's summary of its impressive 200 year history. So why is that? Is Colgate too embarrassed by its fanciful foray into food or is this perhaps just another fake news story which has managed to fool the media and marketing industry for years?

Colgate Kitchen Entrees are certainly frequently cited as a text book example of how brand extensions can fail. I have my views – but let's not allow the truth to stand in the way of a good story!

It would be great if Colgate was to clarify once and for all, though.

In the meantime, Colgate Lasagne – with the white sauce oozing out so deliciously reminiscent of toothpaste – is this week's nominee for The Octopus TV Failure® Awards.

Yum Yum - tongue tinglingly tasty!

April Fools' Special

April Fools!

I love April Fool's Day!

Every year some of the most creative minds flex their intellectual muscle to come up with cleverer and cleverer ways to fool us, and each year we lust for more.

We know it is coming and yet still we are fooled like goldfish with our 3 second memory span – how was that ever tested??

We are always told that history repeats itself and yet we still fail to learn the lessons.

This year they have decided to remove the word "gullible" from the dictionary.

We rejoice in our gullibility. I am compiling a list of favourites over the years … watch this space.

In the meantime, here is a flavour of some of my favourites.

Spaghetti Trees

This brilliant report appeared in the highly respected current affairs programme, *Panorama*, in 1957.

Richard Dimbleby talked about the fantastic weather leading to an "an exceptionally heavy spaghetti crop" in the Canton of Ticino, the southernmost canton of Switzerland.

With the usual gravitas of BBC news reports from the era, we were told that another reason why it might be a bumper year was due to the disappearance of the retched spaghetti weevil.

At a time when spaghetti was little known in the UK the general public swallowed it whole.

Viewers contacted the programme to ask how they themselves could grow their own spaghetti trees. They were apparently told to stick some spaghetti in a tin of chopped tomatoes and hope for the best!

CNN called this "*the biggest hoax that any reputable news establishment ever pulle*d."

Saving Face – Big Ben Becomes Digital Dave

The BBC, that pillar of British broadcasting, loves its annual permission to play.

In 1980 what was then the BBC Overseas Service and later became the BBC World Service tried to convince people that Big Ben was going to go digital with the new clock emitting "beeps" instead of "bongs" on the hour.

The BBC went on to point out that the first people to contact them could win the iconic hands.

BBC apologised for weeks afterwards to the several people who believed the report.

The same prank ran again in April 2015 in the Daily Star who reported, "Due to dwindling tourism in the capital Big Ben's analogue clock face will become digital. Foreign tourists have reportedly been complaining about the "olde worlde" feel of London and want to bring it up to date with other cities such as Tokyo, Beijing and Moscow. Many say they cannot read the time on Big Ben and in a survey 90% of respondents said they would only visit the 156-year-old attraction if it were more modern."

The *Daily Star* reported that some Londoners expressed shock and outrage at the move, "Fruit and veg seller Reg Roberts, who has worked on his stall on the corner of Westminster Abbey for almost 60 years, said, 'I can't believe they would do this. I blame the EU.'"

Cabbie Lance Smears added, "How much is this going to cost? It goes without saying it will be us, the taxpayer, footing the bill."

Taco Liberty Bell

We are very precious about our National Monuments.

Outrage about Big Ben going digital was predictable. Rinse and repeat!

Similarly, the public burst a blood vessel when on 1st April 1996 Taco Bell took out an ad in various newspapers saying that it had bought the Liberty Bell and was renaming it the "Taco Liberty Bell."
Blind to the date of the announcement, people were predictably outraged and several called the National Historic Park in Philadelphia where the bell is housed to express their anger.

White House press secretary, Mike McCurry, when asked about the sale to Taco Bell brilliantly explained that the Lincoln Memorial had also been sold and would now be known as the "Ford Lincoln Mercury Memorial."

Toucan Play At That Game

Guinness are mischievous masters of the April Fool.

As we have seen mega media mischief makers play with us every year.

Real joy is experienced when the roles are reversed and serious major publications fall for the joke. The mainstream media becomes the lame stream media.

And so it was in 1998 when the Financial Times ran a story that a sponsorship deal had been struck between the Old Royal Observatory in Greenwich and the Guinness brewery. According to the report, Guinness was to be the official beer sponsor of the Observatory's millennium celebrations. Under the deal Greenwich Mean Time would be renamed "Guinness Mean Time." Instead of counting seconds in "pips" the Observatory would count them in "pint drips."

The FT criticized Guinness saying it was setting a "brash tone for the millennium."

Red Faces At The Pink Paper

There were red faces at the pink FT when they realised they had been duped. Publishing a retraction, the FT pointed out that the news "was apparently intended as part of an April 1 spoof."

Food For Fools

There have also been several great food and beverage related April Fool's gags in addition to *Panorama*'s Spaghetti Trees.

In 1998 Burger King took a full-page advertisement in *USA Today* announcing the introduction of the "Left-Handed Whopper." This was to be introduced especially for the 32 million left-handed Americans.

According to the ad, the new left-handed whopper included "the same ingredients as the original Whopper (lettuce, tomato, hamburger patty, etc.). However, the left-handed whopper had "all condiments rotated 180 degrees, thereby redistributing the weight of the sandwich so that the bulk of the condiments will skew to the left, thereby reducing the amount of lettuce and other toppings from spilling out the right side of the burger. Finally, after years of neglect, left-handed eaters will no longer need to conform to traditional right-handed eating methods when enjoying America's favourite burger."

Fact From Fiction

In this day and age it is hard to separate fact from fiction. We live in extraordinary times. Reality is far more bizarre that fiction.

The 2011 film *Contagion* has a stellar cast including Gwyneth Paltrow, Matt Damon, Jude Law, Elliott Gould and Laurence Fishburne in the midst of a pandemic as the CDC works to find a cure. WHO? (See what I did there??!)

CDC is The Center for Disease Control and Prevention, the leading national public health institute of the United States.

As Coronavirus / Covid-19 takes its stranglehold on our lives and temporarily destroys our economy *Contagion* looks more and more like a documentary with a virus which allegedly started with bats in China contaminating the food supply.

The movie is peppered with phrases which now punctuate the incessant news coverage:
"Social distancing"
"Wash your hands"
"Stay at home"
"Underlying medical conditions"
"Bats! "
"BATTY!"

Science fiction is science fact.

Baby Food For Adults

In light of the brilliant food fool-yous, you'd be forgiven for dismissing baby food for adults as a joke. Only it was not announced on 1st April nor was it meant as a joke. Which is why it is this week's Octopus TV Failure Awards nominee.

Gerber

Gerber Products Company is a leader in early childhood nutrition. Gerber was founded in 1928 in Fremont, Michigan which is about 35 miles southeast of Hart, home of the Silver Lake Sand Dunes. According to the website, it is "nestled along the shore of Lake Michigan and Silver Lake you will find Michigan's great outdoors with massive sand dunes, beach buggies and miles of hiking and walking trails. Silver Lake Sand Dunes are blessed with several miles of pristine Lake Michigan shoreline perfect for sunning and swimming."

"The Gerber story, by most accounts, began in 1928 on the production lines of the Fremont Canning Company. Gerber has continued to grow throughout the years. Nearly 190 food products are now labelled in 16 languages and distributed to 80 countries. Always dedicated to the health of its young consumers, Gerber has maintained one of the world's largest private research facilities dedicated exclusively to infant nutrition."

Go Gerber!

Gerber joined the Nestlé family on September 1, 2007. As Gerber proudly proclaim, "...at Gerber, research informs everything we do – from the products we make, the nutrition education we deliver and the services we offer. Gerber provides resources from the Feeding

Infants and Toddlers Study (FITS) for health care professionals at Medical.Gerber.com/FITS and for parents at Gerber.com."

Let's face it, locked away in social isolation we start to experiment. Those unlabelled cans lurking threateningly in the darkest corner of the cupboard – undisturbed for years – get opened and contents devoured. Even if Rufus the dog misses out on what was supposed to be his. Even baby food looks tempting and can be comforting. Who doesn't like the occasional mushed banana?

But launching a whole product range is a step too far.

In 1974 Gerber, nevertheless, thought it would be a good idea to release Gerber Singles - small servings of food targeted at college students and single adults, packaged in jars that looked the same as those used for baby food.

The range included Beef Burgundy, Creamed Beef and Blueberry Delight.

Gerber Singles made *Business Insider*'s list of the top 11 Biggest Food Flops of all time. Who needs 10 when you can turn the volume up to 11. Tap the Spinal!

As *Business Insider* pointed out, Gerber Singles had a "**Fatal Flaw:** As it turned out, pre-portioned packages of meat mush didn't exactly scream "cool" to young singles.

The epic flop is one of the most frequently referenced brand failures of all time."

Gerber's Singles also made it in *Mental Floss*'s 10 Food Products That (Thankfully) Flopped pointing out that, "It didn't take long for Gerber execs to figure out that most consumers, unless they were less than a year old, couldn't get used to eating a pureed meal out of a jar—particularly one depressingly labelled "Singles."
"Baby food for grown-ups was pulled from the marketplace shortly after its birth."

Single – Ready To Mingle

Being single is a choice but not everyone's first choice. By calling the product 'Singles,' Gerber shone what, for some, was an unwelcome spotlight on the fact that people were romantically a failure. According to Susan Casey from *Business 2.0*, "They might as well have called it 'I Live Alone and Eat My Meals from a Jar.'"

The Gerber brand is inextricably linked to baby food in the same way as Colgate is inextricably linked with health and that fresh minty taste.

For those who love their meals mushed and wanted to be reminded that they were single, Gerber's Singles was the perfect product. For the rest of us, it was a brand extension too far and is accordingly is this week's Octopus TV Failure® Awards Nominee.

Coronials

Whilst baby food for adults was a failure, the demand for baby food for babies is likely to increase. With so many forced to stay at home, expect to see a baby boom in the next 9 months with the generation of Quarantinies / Coronials. You heard it here first, folks!

Now time to toddle back to explore what else is lurking in the bowels of the kitchen cupboards.

Bon Appetit!

Delorean DMC-12
NOVEMBER 2019

Managing The Message!

What a week for again demonstrating the power and importance of getting the marketing message right.

Pelaton, the exercise bike firm, was told to get on its bike and had US$1.5 billion wiped off its value after a backlash against its Christmas advert which was seen as "sexist and dystopian."

At the same time a local family run hardware shop in Wales was widely hailed as having produced the "Christmas ad of the year" on a budget of only £100!

With only one week away, election fever is throbbing and my 100% record of predicting human behaviour, the media coverage and results remains intact.
From products to politics, managing the message is key.

NATO @ 70 – 2 Right!

It was great to see World Leaders – those master media message manipulators - in town for the 70th anniversary of NATO, the organisation that "protects nearly a billion people."

Justin Time – 2 Faced / 2 Per Cent / 2 Fingers

The constant media coverage provided a perfect reminder to be careful what you say as inevitably what you do say – especially in those unguarded moments - will get back to the person you least want to hear it!

Yet, those in the public eye never seem to learn or do they?

In an age where every step you take and every breath you take, the media will be watching you
(Sting! Don't get stung!) It is perhaps as credible as an alibi in Pizza Express that politicians don't know exactly what they are doing and so we have headlines of world leaders caught on camera apparently mocking Trump.

"You just watched his team's jaws drop to the floor!" said Justin Trudeau, Prime Minister of Canada.

In retaliation, Trump called Trudeau, "…two-faced…" suggesting that the comments from Trudeau were a result of US demands that Canada should meet NATO's 2% defence spending target. Trump pointed out that Trudeau is "…not paying 2% and he should be paying 2%."

2 faced?
2 percent?

2 fingers to Trudeau from Teflon Trump!

Mega Motorcade

During his time in London President Trump was again my neighbour at Winfield House here in Regent's Park. There are 12 acres of ground, the second-largest private garden in London after that of Buckingham Palace. Winfield House has been the official residence of the United States Ambassador since 1955.

During Presidential visits the skies above my home are punctuated with helicopters and the mega motorcade parades around the park. As one world leader told me, once you've been in a motorcade you can never go back.

Which is a roundabout way (geddit?) to bring us to this week's nomination for The Octopus TV Failure® Awards – where the car is the star and Alec Baldwin is the link.

This week we look at an icon of the 80's and star of *Back To The Future*, the DeLorean DMC-12. A tale of sex, drugs, fraud and gull wings.

First Love

We all remember our first car. Mine was a Lotus Elite. The only reason it had a heated rear window was to keep your hands warm whilst you pushed it. One of the unintentional features were the pop up lights which would alternate going up and down as you drove along...and yet through rose-tinted teenage glasses it looked great and made me feel fantastic!

The cassette machine - which could also be used for dictation - blared out music from my childhood icons many of whom are now friends and business partners.

The Lotus Elite was the first step in Colin Chapman's plan to take his Lotus company upmarket.

From 1979 until his death of a heart attack in December 1982 Colin Chapman was involved with John DeLorean in the creation of the DeLorean DMC-12.

At 40 John Z. DeLorean was the youngest ever Vice President of General Motors. DeLorean was a brilliant engineer. He was also a superb salesman with a flair for self-promotion. Another master of the media message. He himself proudly boasted in 1996, "I don't think there's a car running today that doesn't have something I created."

Financing The Future

With his Californian tan, plastic surgery enhanced jaw and reputation as a playboy genius, DeLorean successfully raised $10m from 343 dealers in exchange for shares in his new business, in spite of having nothing tangible to sell.

Further money came from the British taxpayer. Jim Callaghan's Labour Government provided £54 million in grants and loans, no doubt beguiled by the promise of a state-of-the-art 72-acre factory in the Belfast suburb of Dunmurry and the creation of 2600 jobs.

The DeLorean car brought pride to a troubled Ulster and helped to bring "jobs, homes and hope" to the region. DeLorean also attracted a number of high profile investors including Johnny Carson and Sammy Davis Jr.

DMC-12 was designed by Italian automobile designer Giorgetto Giugiaro and engineered by Colin Chapman and Lotus Cars.

Development was completed in 25 months. The DMC-12 had stainless steel body panels, a rear-mounted 2.85 litre V-6 PRV (Peugeot, Renault, Volvo) engine, and gullwing doors.

The first DeLorean rolled off the assembly line in January 1981. Production began and ended in the early '80s after making about 9,000 cars. The company had projected that break-even would be somewhere between 10,000 and 12,000 cars.

Run DMC Run

With the marketing slogan "To live the dream" the car looked divine. Rather than a dream in reality the car was a nightmare.

Car assembly was haphazard. The beginning was rocky. Deadlines were missed. DeLorean asked for more government money and the first cars had to be reworked in the US.

The reason so many photos of the DeLorean have the gull-wing doors open is probably because the doors would not close! Often the doors did not fit properly and got stuck.

The dye from the floor mats rubbed off onto passengers' shoes. The battery lasted about as long as Donald Trump's members of staff. The first time US chat show king Johnny Carson took the DeLorean out it broke down.

The stainless steel panels showed fingerprints and stained easily.

It was also horrendously under-powered. 0-60 mph in a miserable 10.5 seconds. As one critic pointed out "Objectively speaking, the DeLorean is slow, it doesn't really handle and there's more entertainment to be had out of monitoring the reaction to it than actually driving it. But I just can't help loving the thing."

Initially, the car sold well even outperforming the Porsche 911 in the last quarter of 1981. But soon reality hit and unsold DeLoreans began piling up at Belfast's docks and the company ran out of money.

Many different factors contributed to DeLorean's failure to survive: poor sales, high cost overruns, unfavourable exchange rates, an extreme winter (people do not buy cars when there is snow on the ground) and not least the extravagant lifestyle of John DeLorean.

The company went bankrupt in 1982. Over 2,000 people lost their jobs and investments of over $100 million disappeared.

White Lines

On 19 October 1982, DeLorean was arrested in the LA Sheraton Plaza Hotel having allegedly taken part in a plot to smuggle 100kg of cocaine.

DeLorean was charged on eight counts of conspiring to possess and distribute cocaine.
Following a 62-day trial an LA jury acquitted DeLorean based on his defence that it was FBI entrapment.

Grand Theft Auto where the car's the star

DeLorean's troubles were not over.

The Official Receiver found evidence suggesting that DeLorean had conned several investors as well as the British Government and that money had been diverted to DeLorean's personal accounts.

About £10m had gone missing. Some of the money that was to be used by Lotus Cars for research and development work for the DeLorean company was in fact siphoned away via a Swiss-based company. Colin Chapman died before he could answer the accusations.

DeLorean, whilst accused of defrauding the UK government, was never prosecuted. DeLorean insisted on his innocence, 'I've made mistakes in my life, I admit. But I've never done anything dishonest.'

In 1999 John DeLorean - previously estimated to be worth US$170m - was forced to declare personal bankruptcy and faced numerous lawsuits. He lost everything including his 400 acre estate in New Jersey, later converted to a golf course by Donald Trump.

John DeLorean died of a stroke in 2005, aged 80. His gravestone at Michigan's White Chapel Memorial Cemetery depicts a DMC-12 with its gull-wing doors open. Even in death it was impossible to close the doors!

In a 1996 TV interview, John DeLorean pointed out, "I ended up living a lifestyle I couldn't believe. The tragic thing is, you start to believe your own press. I fully confess to becoming egomaniacal. You believe you are omnipotent, and you are surrounded by people who tell you what you want to hear rather than the truth. Never judge a man unless you've walked a mile in his moccasins."

The Guardian's obituary described him as an 'American car-maker and conman.'

Back To The Failure / Forward To The Past

The dream of the DeLorean is, however, not dead.

There are still over 6,000 original DMC-12s in the world and following a change in the legislation regarding low volume manufacturers, there are plans to restart production.

The Company is now run out of Texas with Stephen Wynne as its CEO. Wynne is looking at producing a new DeLorean, possibly an electric vehicle or SUV.

"The brand strength is there," Wynne said, "The first decade was drug deals and bankruptcy, the second decade was people thinking it was the movie prop car. But now it's back to people thinking of it as a good, solid classic car that was very innovative -- and very desirable."

...and it's not just the roads that may again be dotted with DeLoreans...

Where They're Going You Don't Need Roads

Paul DeLorean, nephew of John DeLorean, CEO and chief designer of DeLorean Aerospace, is on a mission to "bring the freedom and exhilaration of personal air transportation to the masses "with a real life flying car. The DR 7 is built like a Formula 1 car for the sky!

Where they're going they don't need roads.

TOFA

The DeLorean story is itself a fascinating fable of failure and fraud. Parliament's Public Accounts Committee branded the project as "one of the gravest cases of the misuse of public resources to come before us in many years."

The DMC-12 was an iconic symbol of the 80s and gained an almost cult like following thanks to the *Back To The Future* movies released a number of years after the company's demise.

In reality, without the benefit of a plutonium-fuelled nuclear reactor and Flux Capacitor it was a nightmare of a car: a badly engineered gimmick and a commercial disaster with mythic significance akin to Belfast's other globally famous failure, RMS Titanic.

88 mph Hmmm...

Delorean Trumped – Brilliant Baldwin

So what has this got to do with Alec Baldwin?

Well, the DeLorean story is back in the news with the release this year of XYZ Films' documentary "Framing John DeLorean" starring none other than Alec Baldwin as DeLorean.

And it is the very same Alec Baldwin who continues to delight the world with his brilliant portrayal of President Teflon Trump on Saturday Night Live.

So Trump is back in Washington as Congress launches the next stage of impeachment. Just don't expect a DeLorean to be in the motorcade.

TWITTER PEAK
TRUMP IMPEACHED
21 YEARS AFTER CLINTON
PUBLISHED DECEMBER 2019

#Impeachment

As I predicted, headlines around the world are today screaming the fact that The House of Representatives has voted to make Donald Trump only the third President in US history to be impeached.

It is indeed exactly 21 years to the day when Bill Clinton was impeached on December 19, 1998, by the House of Representatives on grounds of perjury to a grand jury.

Prediction – Teflon Trump

Trump will now face a trial in the US Senate.

I have already predicted, however, that Teflon Trump will survive that trial, the actions of the Democrats will backfire and Trump will become even more powerful securing a second term! [A buoyant economy would have ensured victory. The effects of Covid-19 may have an adverse effect on what would have been a sure-fire bet.]

If only they had listened ….

Throughout all of this, expect the Twittersphere to continue to explode.

#Hashtag

Certainly Twitter is Donald Trump's communication weapon of choice. In fact, Monness Crespi Hardt analyst James Cakmak believes if @RealDonaldTrump were to leave Twitter, the social media company would lose as much as $2bn in market value. "There is no better free advertising in the world than the President of the United States," said Cakmak.

It's fair to say, however, that not even @POTUS @RealDonaldTrump would be tempted by this week's nominee for The Octopus TV Failure® Awards – The Twitter Peek.

Twibel – Be Careful What You Say

Historically, many defamation cases have concerned broadcasters and newspapers. Today we are seeing more and more involving social media.

One of the first big Twitter Libel ("Twibel") cases was taken by the Conservative peer Lord McAlpine against Sally Bercow, the wife of the former House of Commons speaker, John Bercow. We have also seen Katie Hopkins held to task for her Tweets.

Elon Musk, the Tesla founder, was in the news last week having won the US$190m defamation suit against him over his Tweet.

I predict that the ease with which people can post messages via social media will give rise to a significant increase in the number of defamation cases.

Be careful what you say!

NOTE

Yet again my prediction came true.

As I predicted, Trump was impeached by The House and acquitted in the Senate trial.

:Cuecat
PESTILENCE, PREJUDICE & PARASITE

As pestilence, plague and prejudice persist in populating the press it is refreshing to see such positive praise for *Parasite*.

Bong Joon-Ho's brilliant South Korean satire made Oscar history by being the first film not in the English language to win Best Picture after taking Best Director, Best International Film and Best Original Screenplay.

The Oscar ceremony itself was punctuated with references to the controversy around lack of diversity starting with Chris Rock pointing out what was missing from this year's director nominees: "vaginas."

Rebel – with some claws – Wilson pointed out at the BAFTAs, there was prejudice in all of this year's Awards as "felines" failed to receive any nominations.

Rebel's film was, however, nominated for an Octopus TV Failure® Award ...cue *Cats*!

An elegant and effortless link to this week's Octopus TV Failure Award® nominee - :CueCat

Hello Kitty!

This week a letter flooded in from our reader, Derek Walker: "Andrew Eborn, Please tell me that the ":CueCat" will make an appearance on this list. I had a front row seat for this train wreck! Oh the horror of this idea."

Creative juices oozed from every orifice during the dot com interweb boom. The "new economy" hype gave rise to some glorious products and unprecedented failures.

Investors, not wanting to miss out on the next big thing, threw their hard earned cash in support of some of the most deliciously bizarre products.

:CueCat, - with the purrfectly pretentious colon at the start of its name - is a prime example. A poster pussy with peculiar punctuation.

:CueCat was invented by the wonderfully named Jeffry Jovan Philyaw aka J. Hutton Pulitzer, "one of the most prolific independent inventors of modern times."

Launched in 2000 by Digital Convergence Corporation, :CueCat is a cat shaped bar code scanner which enabled users to scan a special bar code ("cue") which would then shepherd them effortlessly to the relevant webpage. This avoided all the hassle of having to type long web addresses. It was the bridge between the physical world of the printed page and the internet. The ambition was to make the :CueCat barcode the standard for advertising. A cat designed to work side by side with your mouse.

Money In The Kitty

Funding flooded in from several heavy hitters. Derek Walker told me, "I was at RadioShack when they wrote a $30 million check for that thing. $30 million. I get sick thinking about that."

Radio Shack were not alone in putting money in the kitty. Many fat cats were wooed and wowed by the sales skills of JJP. Belo Corporation, the parent company of the *Dallas Morning News* and owner of a number of TV stations, shelled out US$37.5m, Young & Rubicam $28m and Coca-Cola $10m.

In total $185 million was secured. Juicero would be proud!

Catalogues

:CueCat was made available free of charge via many outlets including Radio Shack and was also sent unsolicited to various mailing lists such as subscribers to Wired and Forbes magazines.

:CueCat's investors supported the roll out and :CueCat bar codes started to appear in a number of publications such as the *Dallas Morning News* and other Belo owned publications, as well as in product catalogues for Radio Shack and others.

In addition, audio tones in programmes and commercials could act as web address short cuts, provided your TV was connected to a computer and both were switched on.

Bad Kitty

Derision flowed as freely as Trump's tweets, as critics got their claws into :CueCat.

Walt Mossberg, the principal tech columnist for the *Wall Street Journal* wrote, "On the first standard, convenience, the CueCat fails miserably...In order to scan in codes from magazines and newspapers, you have to be reading them in front of your PC. That's unnatural and ridiculous."

Jeff Salkowski wrote in the *Chicago Tribune*, "You have to wonder about a business plan based on the notion that people want to interact with a soda can."

Glittering prizes were showered upon :CueCat.

:CueCat topped *Gizmodo*'s list of worst gadgets of the decade and *PC World* included it as one of the worst 25 products of all time.

Pussy Footing Around Privacy

There were also concerns around privacy. Each :CueCat had a unique serial number and users had to register their ZIP code, gender, and email address. As a result, personal profiles could be formed, and content could be tailored accordingly. The Denver-based Privacy Foundation expressed concern that a user's reading and television habits could be tracked with precision, stored in a central data bank and resold.

Paranoia over privacy was accentuated by the fact that Digital Convergence registered the domain "digitaldemographics.com."

Clearly Digital Convergence's business plan was founded on profiting from its marketing data base.

Michael Garin, the company's president, insisted however, "We cannot, we do not, and we will not track individual information."

Hacked Off

Reports of hacks and cyber-attacks are now almost daily. Way before NHS, WPP, Yahoo and others came to our attention, Digital Convergence was a high profile victim.

To add to its woes, the feline company suffered a security leak when a tech employee left with a development computer and connected to an unsecured net connection and - surprise, surprise - was hacked. About 140,000 :CueCat users had their personal data stolen including their name, email address, age range, gender and zip code. What was the price of privacy? Digital Convergence offered each victim a $10 gift certificate to Radio Shack. Hmmm...

De-Clawing The :Cuecat

The :CueCat was easy to reverse-engineer. A number of sites provided details of how to neuter the cat.

Curiosity Killed The Cat

As a result, several hackers modified the :CueCat for their own purposes such as cataloguing book or CD collections. The problem for Digital Convergence was that the re-engineering could bypass the company's marketing database, the company's lifeblood.

Digital Convergence got tough and threatened legal action asserting that users did not own the devices and had no right to reverse engineer them. Digital Convergence also tried to change the terms of its licence to prevent reverse engineering.

The hackers were hacked off.

Cat Fight

Interesting questions for us lawyers arose about whether people could take apart freely distributed hardware and write separate software applications. The right to "reverse engineer" hardware for certain purposes has been well-protected in a number of jurisdictions.

There were also questions raised about "enforceability" where a company attempts to assert that a licence is triggered simply by using its hardware.

The old sales strategy of giving away razors to sell razor blades may not always work in the tech space where you give away hardware and try to make money via the software. With software you can make your own blades.

Cue The End

In the end :CueCat was a miserable feline failure and investors lost their US$185million. The database servers which provided the code to Internet URL linking ceased operation in January 2002 and the desktop client is no longer supported.

Interestingly, :CueCat is worth more dead than alive. Originally given away for free, you can still buy :CueCat via your favourite online shopping site, some for more than $19.99!

Unleashing The Cat

Making it easier for consumers to buy whilst building a valuable database makes sound commercial sense. Today we take for granted the ubiquitous QR codes that can be read easily with smartphones.

Smartphones do not need to be plugged into a PC and they can obviously do far more than just scan QR codes.

Just like the other Octopus TV Failure® Nominee Twitter Peek, no need to pony up for one trick tech ponies.

A Catalogue Of Failures

:CueCat had everything:

- pretentious branding - :CueCat with a colon;
- large-scale attempt at aggregating user data to build profiles and then profit by selling that data to advertisers;

- major security vulnerability leading to the risk of exposure of the data harvested; and
- threatening customers who dared to use the :CueCat for anything other than its originally intended purpose with the power of copyright law.

As one critic put it, ":CueCat appears to us as a mirage, a failure, a loser, but unlike all the other crap gadgets, the future it dreamed of came into being. From within that abyss every sleazily obvious effort to hide consumption in convenience, every privacy failure, every data breach, and every DMCA takedown, :CueCat is smirking back at us."

:CueCat was a commercial failure. It was difficult to use reliant on tethering to your computer and there were concerns regarding privacy. Talk of a wireless version was too late.

For all these reasons :CueCat is this week's nomination for The Octopus TV Failure® Awards. Purrfect!

FINAL THOUGHTS

Learning The Lessons Of History

As I always point out, history repeats itself. We could, and should, learn lessons from history and yet we fail to do so. Perhaps our biggest failure.

Human behaviour is predictable which is why I have so successfully predicted major events and the results of elections as well as the media coverage with what many have said is "alarming accuracy."

What is clear is that we will fail and fail again.

In shining a light on these failures it is my mission to encourage everyone to embrace and learn from their mistakes and the mistakes of others.

We are all human.

If you prick us, do we not bleed? If you tickle us, do we not laugh? We need fewer pricks and more tickles!

We all have dreams and aspirations.

We all have fears and insecurities.

We must not let those fears cripple opportunities.

As Aristotle pointed out between football matches, to avoid criticism, "say nothing, do nothing, be nothing."

Communication is key.

So often perception and reality are not aligned. People need the courage to speak out. Silence merely serves to condone the unacceptable.

People should have the intelligence to question everything including the motives of those who seek to spread damaging misinformation without raising issues directly with the person concerned.

I always try to understand everyone else's point of view and how prejudices may have shaped those views.

Everyone deserves respect. Everyone deserves the opportunity to have perception explained and the opportunity to clarify any misunderstandings.

We should all strive to shed more light and less heat on situations.

To avoid regrets, do today what needs to be done:

- Finish that unfinished business!
- Clarify that misunderstanding!
- Build that bridge!
- Look up!

- Make that phone call!
- Say how you feel!
- Say sorry!
- Say I Love You!
- Today may be your last opportunity...
- Carpe diem! (Go fishing TODAY!)

"From failed products and services to campaigns and ads we would rather forget, I want to encourage organisations and brands to be better at learning from failures not just ignoring them and pretending they never happened."

Send your nominations now together with full description and IP cleared images to TOFA@Octopustv.com

Every day at 07:07 Octopus TV broadcasts more of my exclusive interviews with great guests from around the world:
- from Australia to Zambia
- from all walks of life including major public figures and celebrities, legends of sports, music, film, theatre, titans of industry as well as everyday heroes
- learn how to be the best you you can be
- learn to cook with top chefs
- learn from the world's experts from chess to wrestling, from perfect pastry to paper planes
- lifestyle - yoga, meditation

-   exercise your brain with Afternoon Tease, brain teasers, puzzles, mind bogglers

and much, much more

Hundreds of hours of fantastic content available for licensing.

BREAKING NEWS

At 0606 on 0606 Andrew Eborn announced the launch of EBORN NEWS NETWORK (ENN)

We are drowning in a sea of information – 99.9% of which is rubbish. Anyone, anywhere and at any time can be a self-professed expert. There is evidence to support any view – however outrageous.

The news is dominated by soundbites from those Sultans of Spin.

In this post-truth age it has never been more important to have evidenced-based journalism.

As Jimmy Wales, the co-founder of Wikipedia, pointed out, "The news is broken and we can fix it."

QUESTION EVERYTHING

REALISE REAL LIES

We will bring you stories directly sourced & evidenced based from around the world.

If there are particular stories you feel should be subjected to a pressure test to find out whether they really stand up to serious scrutiny, if you want to avoid the predictable errors and omissions of others, if you want to be a guest on the Andrew Eborn Show or if you want to engage me as an inspirational presenter, broadcaster and speaker get in touch.

Andrew.Eborn@OctopusTV.com
@OctopusTV
@AndrewEborn

WHAT OTHERS SAY
ABOUT ANDREW EBORN

Andrew Eborn is the founder of *Canned Laughter -It's OK Not To Be OK* promoting mental health and wealth - as well as Octopus TV's *Head to Toe of Health* with the leading medical institutions around the world.

"Andrew Eborn is the Joe Wicks of The Mind and Mental Health"

UKTI Rob Walker - "Andrew Eborn is an inspirational ambassador for the technology, media and sport sectors and a powerful advocate for British business."

"Magical" - Sir David Frost

"A pleasure" - Harvey Goldsmith CBE

"Andrew Eborn is a very talented interviewer. Andrew is a Barrister, magician, a true showman & President of Octopus Television, London. When Andrew interviews his subject he makes them feel extremely relaxed & a joy to work with." - Linda Rifkind

Andrew Eborn is "an extraordinary interviewer" - James Fortune, Hollywood Director

"Brilliant" - The Magic Circular

"Andrew Eborn has now interviewed me twice. He's very good at it too because he asks tough questions without being tough and, like Sir Michael Parkinson, he asks questions the viewer would like to ask." - Paul Gordon

"If it wasn't for Andrew Eborn I would not have had the power to do what I am doing. ... " - Charmaine Davis

"I like your interview style. It's buzzy, prepared and strong. You do it very well."

"Brilliant and buzzing" - Ed Hood

© Octopus TV Ltd 2020.

About the Author

Andrew Eborn, President of Octopus TV Ltd and Knot The Truth Ltd, is a renowned international lawyer, strategic business adviser, producer, broadcaster, writer, columnist, futurologist and in high demand as a presenter, magician and inspirational speaker around the world.

He is now working across the IP value chain including the creation and licensing of content in all media from production, publishing, distribution, supply of talent, management, promotion, immersive technology, holograms and AI.

For many years Andrew has empowered companies to face the challenges of changing markets, maximise the return on their rights and assist with the strategic development of their businesses.

In his series, Andrew shines a light on the products and services, brand extensions and campaigns that failed to take off and have as a result earned entry into The Octopus TV Failure® Awards.

As Andrew points out, "We always celebrate success whilst hiding the failures that led to that success.

The Octopus TV Failure® Awards finally give Failure® the attention it deserves.

If necessity is the mother of invention, Failure® is the father of success.

From failed products and services to campaigns and ads we would rather forget, we want to encourage organisations and brands to be better at learning from failures not just ignoring them and pretending they never happened."

© Andrew Eborn 2019 ALL RIGHTS RESERVED
® Failure is a registered trademark owned by Octopus TV Ltd. ALL RIGHTS RESERVED

Email Andrew
Andrew.Eborn@OctopusTV.com

Follow Andrew
Twitter @AndrewEborn and @OctopusTV

www.octopus.tv

THANK YOU

To my amazing mother, Valerie Eborn, and her parents, William & Sylvia Wheeler, without whom I would not be the person I am today.

To my late father, Gerald Eborn, and his parents, Ernest and Lily. Gerald was determined, charming and with a wicked sense of humour. Family occasions still buzz with the re-telling of Gerald's exploits. He lived a life full of love, adventure and a huge dollop of mischievous fun.

It is only by becoming a parent yourself that you can fully appreciate your parents.

Appreciation is always appreciated.

Aristotle — 'Give me a child until he is 7 and I will show you the man.'

Nostalgia ain't what it used to be ...

To the wonderful John & Vivien Eborn, Heather & John Burden, Cecil & Beryl Hibbert, Cookie, Smudge, Moriarty, Rodney, Priscilla, Michael & Daniel Eborn, Deborah, Simon, Ben, Sarah, Clare & Helen Martin, Rebecca "Little Smig", Adrian Glossop, Anthea & Paul Stern, Keith & Elizabeth Burden, Carina, Rosina, Graham, Christopher, Stuart, Angelina, Anita, Miriam, Stefan, Anastasia, Richard, Margaret, all Eborns, Wheelers & Parrs whether now known or hereafter invented ...et al (we all love al !)

To all those people & places that played any part in my life: High Barnet, Bracklesham Bay, North Gate House, Mrs Shoesmith, Mr Wilds, St Anne's, Peter Carmichael, Cubs & Scouts (undefeated Bob-A-Job Champion) Prebendal, Lord Wandsworth College, Durham University, St Chads, Buffalo Head, College of Law, Inner Temple, Sir William Macpherson, Seddon Cripps, Francis Brennan, OJ Kilkenny, U2, Toyah, Hazel O'Connor, Mike Read, Bob Geldof, Sussex Beach, Oscar's Den, Dot & Arthur Moir, Ray Mummery, Alternative Arts, London Palladium, Ainsley Harriot, Paul Boross, LWT, Janie Fraser, Nikki Finch, Thames TV, John Fisher, TVS, Vanessa Chapman, John Morgan, Island Visual Arts, Andy Frain, Stephenson Harwood, Cat Stevens/ Yusuf Islam, TV-AM, Royal Albert Hall, Denton Hall, Willie Robertson, Cameron Markby, Tony Morris, Arc, Matthew Howells, Mitch Levy, Myka England, Equity, The Magic Circle, Ali Bongo, Fuji TV, Nobu Kasai, Tetsuo Hamaguchi, Hidesato Hayashi, Funaki san , Jun Minegishi, Sue Hudson, Sadaishi san, Jerry Leung, Masahiro Shibata, Toru Uehara, Koichi Kaneko, Inamura san, Formula Nippon, Octopus Korea (OK), Albert Song, Younghoon Song, Stride Music, Jacqui Lyons, Kazutoshi Miyaki, Kaz Hori, Hide Son, Dentsu, Yo Hattori, Kaz Iwagami, Shin Okura, Meg Niimura, Kiyoshi Nakamura, Hanaoka san, Jonathan Bodansky, F1, Bernie Ecclestone, Judy Griggs, Fuji Film, Adrian Clarke, David Honey, Peter Bennett, Kevin Langham, Luigi Lavarini, John Gough, David Briggs, Colman Hutchinson, Steve Springford,

Mike Whitehill, Fintan Coyle, Alison Jackson, Reed Midem, Ted Baracos, Lily Ono, Lazar Vukovic, Dwina Gibb, RJ Gibb, all other wonderful clients and precious friends. ..and especially to you checking to see if I have mentioned your name - you are in my very special thoughts and always will be .. how could I ever forget you??!?

(I'm working on the school play mentality that if you have enough people in the cast you are guaranteed an audience!)

Fantastic memories and so many, many more to come.

Finally, a huge THANK YOU to all of you who keep sending me suggestions and inspiration. Keep 'em coming

May we NEVER FAIL to FAIL.

Together we will ensure an EXCESS of SUCCESS

... and I LOVE IT!!!!

WS - #0028 - 240720 - C0 - 210/148/7 - PB - 9781838030216